GEOLOGY ROCKS!

EXPLORE ORES

CHRISTINE PETERSEN

Checkerboard Library

An Imprint of Abdo Publishing
abdobooks.com

ABDOBOOKS.COM

Published by Abdo Publishing, a division of ABDO, PO Box 398166, Minneapolis, Minnesota 55439.

Checkerboard Library™ is a trademark and logo of Abdo Publishing.

Printed in the United States of America, North Mankato, Minnesota
102019
012020

Design: Emily O'Malley, Mighty Media, Inc.
Production: Mighty Media, Inc.
Editor: Jessica Rusick
Cover Photograph: Shutterstock Images
Interior Photographs: Biswaranjan Rout/AP Images, p. 11; FastGlassPhotos/iStockphoto, p. 13; Library of Congress, p. 5; Mighty Media, Inc., pp. 16, 17; Opla/iStockphoto, p. 7; RBFried/iStockphoto, p. 15 (top left); Shutterstock Images, pp. 4, 6, 9, 10, 12, 15 (top right, bottom), 19, 21, 23, 24, 25, 26, 27, 28, 29

Library of Congress Control Number: 2019943215

Publisher's Cataloging-in-Publication Data

Names: Petersen, Christine, author.
Title: Explore ores / by Christine Petersen
Description: Minneapolis, Minnesota : Abdo Publishing, 2020 | Series: Geology rocks! | Includes online resources and index.
Identifiers: ISBN 9781532191732 (lib. bdg.) | ISBN 9781532178467 (ebook)
Subjects: LCSH: Geology--Juvenile literature. | Ores--Juvenile literature. | Mining geology--Juvenile literature. | Mineralogy--Juvenile literature. | Rocks--Classification--Juvenile literature.
Classification: DDC 553--dc23

CONTENTS

SUTTER'S **MILL**

In 1839, John Sutter was one of the few settlers in the California wilderness. He was granted a huge area of land. In time, Sutter needed more lumber for building. So, he asked carpenter James Wilson Marshall to set up a sawmill.

For the mill, Marshall chose a spot along the American River. The mill required a steady source of water to run. So, Marshall's workers dug two deep channels. These would bring water from the river to the mill and back again.

On January 24, 1848, Marshall went out to check on the lower channel. Something in the water caught his eye. "I reached my hand down and picked it up," he later recalled. "It made my heart thump, for I was certain it was gold."

Marshall's discovery triggered the California Gold Rush. In one year, 80,000 people came to California in search of gold.

Marshall first tested the gold by biting it. Pure gold is soft enough to be scratched by teeth.

Without planning it, the workers at Sutter's Mill had struck an ore. An ore is an **aggregation** of one or more **minerals** that can be mined for profit. In Marshall's case, that valuable mineral was gold!

Marshall's gold came from a **placer** deposit. An ancient stream caused it to collect there. Over millions of years, the gold was buried. When Marshall's workers dug into the ground, they brought the gold back to the surface!

A SPECIAL **RECIPE**

Have you ever made chocolate chip cookies? Flour, sugar, eggs, butter, and more get mixed together. Then the chocolate chips are added. After they're stirred in, the dough is shaped into balls and baked.

Imagine taking a close look at a cookie before eating it. Can you spot a few chips? You could easily pick out these goodies. But where have the other ingredients gone? They are mixed together throughout the cookie. Your cookie is like Earth's rocks and **minerals**!

Earth's outer layer is called the crust. It is made of rock. Rock is made of minerals.

Most rocks are made up of more than one kind of mineral. Often, these minerals are well mixed. You'd find it hard to collect just one kind. Yet in some areas, rocks

Like the chips in your cookie, ore minerals are always surrounded by other minerals.

All ore deposits are mineral deposits. But not all mineral deposits are ore deposits. Ore deposits must be profitable to mine.

contain clumps of a specific **mineral**. Like the chips in your cookie, they are easier to locate and mine.

These unusually high amounts of a specific mineral are called mineral deposits. This is a geologic term. Mineral deposits have been discovered around the world. If one can be mined at a profit, it is called an ore deposit. That is an **economic** term.

MINERAL BUILDING **BLOCKS**

At a museum's **mineral** exhibit, you might see examples of gold and quartz. Gold looks like a shiny, yellow blob. Quartz crystals can resemble small, white skyscrapers. You would never mix up these two minerals on a test!

Yet under a microscope, gold and quartz have something in common. Both are made from tiny building blocks called atoms. Atoms make up **elements**.

Scientists have identified more than 100 different elements. Every mineral contains at least one. For example, gold is made up of gold atoms. Many minerals contain more than one element. Quartz has atoms of the elements silicon and oxygen.

Different minerals have different combinations of elements. Scientists have discovered about 4,000 minerals. Only about 100 are considered ore minerals. Most of these are the source of useful or valuable metals.

Gold sometimes occurs in quartz deposits. ▶

MAKING **MINERALS**

Minerals form in several ways. In some cases, the way they form causes them to become concentrated. Then, they may be called mineral deposits.

Some minerals form from cooling **magma**. Magma flows upward from below Earth's crust. It may reach the surface and spread across the ground as lava. Lava cools quickly, which does not give mineral grains much time to grow.

Magma that remains trapped underground cools very slowly. Under these conditions, larger mineral grains may form. In some cases, this slow cooling leads to mineral deposits.

For example, the ore mineral chromite can form deposits in cooling magma.

Chromite (*above*) is a source of chromium. Chromium is combined with other elements to make colorful dyes!

Chromite is mined in such areas as India and South Africa.

Chromite is **dense**. So, it sinks as it forms. This means a lot of chromite ends up in the same place. It is an ore of the metal chromium.

Hematite can be red in color. Its name comes from the Greek word for "blood."

Water can also form **mineral** deposits. Hot **magma** releases some of this water. Other water trickles down from Earth's surface. The water flows deep underground through tiny spaces in rock. Along the way, it collects various **elements** and **minerals**.

To form a deposit, the water must be affected by its surroundings in some way. Water entering an open space in rock may boil, cool, or react with surrounding rock. This causes the water to deposit its load of minerals. Copper deposits often form because of water.

On Earth's surface, rocks are continually worn down. Flowing surface water picks up rock pieces. It breaks them down and concentrates them somewhere else. This leads to **placer** deposits, such as the one at Sutter's Mill.

The first mined gold likely came from a placer deposit.

Lakes and oceans carry minerals too. The minerals settle out of the water to form sediment layers. Deposits of the mineral hematite form this way. Hematite is an important iron ore.

AMAZING **METALS**

Most ore minerals are mined for important metals. Metals are shiny and can be shaped easily. They conduct heat and electricity. Metals can be strong and long lasting, especially when combined.

Metals are found in everyday objects. Do you have a bicycle? It may contain the lightweight metal titanium. Copper helps computers run. Try to find other examples in your home and school. All of these metals came from ores.

The metals we use fit into two classes. They can be abundant or scarce. The five abundant metals are aluminum, iron, magnesium, manganese, and titanium. They are found in many common rocks. All other metals are scarce. They include copper, lead, and gold.

To meet the need for metal, huge amounts of ore must be mined each year. In 2018, about 54 million tons (49 million t) of iron ore were mined in the United States. That's about the weight of 148 Empire State Buildings! Yet, it was only 2 percent of the iron ore produced worldwide.

Metals are part of our everyday lives. ▶

TRY THIS AT HOME: MINE YOUR CEREAL!

WHAT YOU'LL NEED

- 1 cup iron-**fortified** cereal flakes
- resealable plastic bag
- rolling pin
- bowl
- 2 cups hot water
- wooden spoon
- long magnet (not black)

WHAT YOU'LL DO

1. Put the cereal in the plastic bag and seal it.
2. Use the rolling pin to crush the cereal.
3. Pour the crushed cereal into the bowl.
4. Add the hot water to the bowl.
5. Stir the mixture until the cereal breaks down. This will take a few minutes. The cereal will get very mushy!
6. Then stir the mixture with the magnet for a couple of minutes.
7. Pull the magnet out of the cereal mixture. Look carefully. Are there little black bits stuck to the magnet? That is the iron from your cereal!

2
4
6
7
IRON

CLASSIFYING **ORES**

Minerals are grouped based on the elements they contain. Of the 11 groups, five contain preferred ore minerals. They are native metals, silicates, carbonates, sulfides, and oxides and hydroxides.

Gold and platinum often occur in their native state. That means these metals are not combined with other elements. Silver and copper may also occur this way.

Silicate minerals contain the elements silicon and oxygen, and often a metal element. This group includes garnierite, an ore of nickel. Nickel can be combined with other metals to make stainless steel. All carbonates contain carbon and oxygen. Rhodochrosite is in this group. As an ore, it is a major source of manganese.

Sulfide minerals combine metals with sulfur. They make up the largest group of ore minerals. Galena is the main ore of lead. Cinnabar is the main source of mercury. Telluride minerals are a special type of sulfide. They each contain the element tellurium and a valuable metal.

Oxides and hydroxides are the last group. Bauxite is the main ore of aluminum. Rutile and ilmenite are sources

Bauxite is formed from reddish clay.

of the metal titanium. This metal is lightweight and resists heat. So, it is often used in airplane construction.

LET'S EXPLORE!

Mining ores is expensive and takes a long time. That's why companies prospect and explore before starting to dig. Prospecting is the search for a **mineral** deposit.

Modern science makes prospecting easier. Special **satellite** cameras find some minerals. Magnets help find minerals such as magnetite, an iron ore.

Once found, a mineral deposit is explored to determine its worth. The goal is to find an ore body. This is a deposit with clear boundaries that contains minerals that can be mined for profit.

Several factors determine whether a deposit is worth mining. One of these is grade. This is a measure of how much of a deposit is valuable minerals and how much is **gangue**. The less gangue there is, the higher the grade. A high-grade deposit is more likely to be profitable.

Before opening a mine, companies also consider how much they can sell the ore for. They compare this with the cost to remove and process it. And, they see if too much of the same metal is already available. If so, it may be wiser to mine it at a later time.

Drilling into a mineral deposit for samples helps geologists determine its grade.

Other factors include how big and what shape the deposit is. The deposit's location also affects how easily and cheaply it can be mined. Exploring helps determine all of these factors.

IN THE PIT

Mining can begin once an ore body has been discovered. One place this has happened is the Chino Mine in New Mexico. From a distance, you might mistake this mine for a canyon.

Looking into the mine, you can tell it is actually the work of humans. The rock has been cut away in layers that dip deep into the earth. This makes it look like a huge stadium. Yet no games will ever be played there. That is because Chino is an open-pit copper mine.

An open-pit copper mine is a busy place! The miners start by blasting away sections of rock to obtain the ore. Large trucks then carry away the broken pieces. The rock must be processed to obtain the copper.

For much copper ore, the first step is to crush the ore. The tiny bits of ore are then soaked in water. **Gangue** falls to the bottom, while copper floats to the top. Next, the metal is melted and separated from any other **minerals**. In the end, nearly pure copper is all that remains.

Open-pit mining is just one type of surface mining. Strip mining is a similar way to remove ores. Miners remove strips of earth one at a time. Waste from each new strip is

The Bingham Canyon Mine in Utah is the largest man-made pit in the world. It is an open-pit copper mine.

left in the old strip next to it. Quarrying is the removal of stone in specific shapes, sizes, and quantities.

DIGGING DEEP

Many ore deposits are found too deep for surface mining. In these cases, underground mines are dug. From there, ores must be carried to the surface for processing.

The world's deepest mines are in South Africa. Gold was discovered there in the Witwatersrand region in the 1880s. The area contains one of the largest gold ore bodies in the world. That makes it a **bonanza**!

Even the biggest bonanza can't last forever. Mining companies in South Africa must dig deeper mines to reach good ores. One mine reaches more than two and a half miles (4 km) underground.

After a mine opens, companies continue to think about profits. They compare the value of the ore to the cost of mining. A mine may close if sale of the ore will be less than the cost to produce it. Or, a mine may close when an ore body is used up.

Many of the world's deepest mines are gold mines.

DANGER!

The US Mine Safety and Health Administration wants you to "Stay Out – Stay Alive." There are mines in every US state. It might seem exciting to play or explore around a mine. But this is very dangerous! It is important to stay away from mine entrances. This is true for active mines as well as those that are no longer in use. No matter how safe a mine might seem, only trained professionals should go near them.

MAJOR METALS ARE FOUND AROUND THE WORLD!

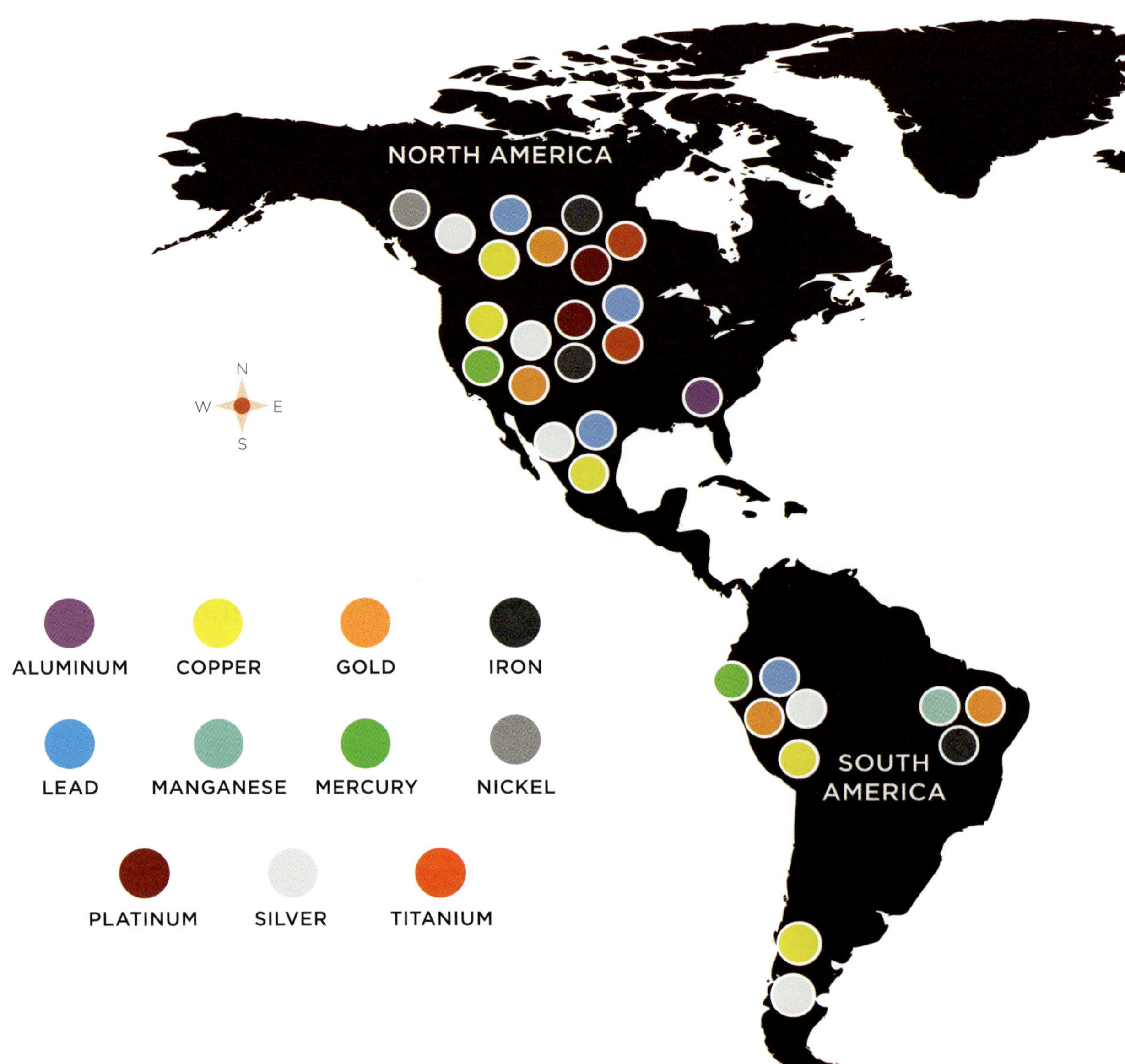

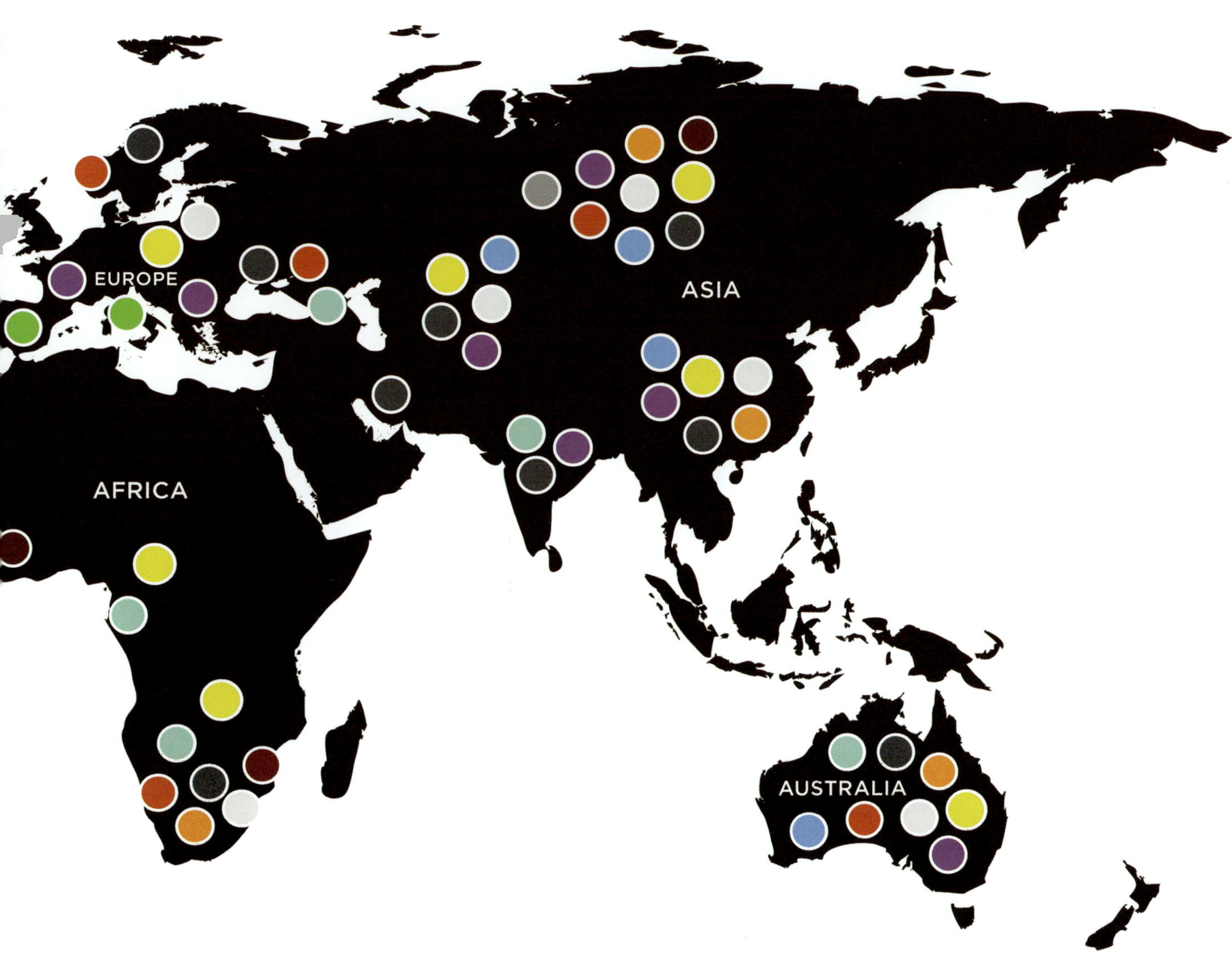
EUROPE
ASIA
AFRICA
AUSTRALIA

ORES IN OUR LIVES

People have been mining minerals for thousands of years. Early on, humans used native metals such as gold and copper. They shaped the metals into practical and beautiful objects. Today, we know how to obtain metals from many different ores.

Ores are one of the most important **natural resources** on Earth. Metals are everywhere in our lives. And, mining provides many different jobs for people around the world.

Yet there is a price for using these resources. Natural **habitats** must be cleared to open some mines. And, mining can cause pollution. So, laws exist to reduce damage to the **environment**.

Mining companies know it is important

A key step in the mining process is restoring mined land. This is called reclamation.

Recycling scrap metal uses less energy than refining metal from ore.

to protect nature and people. One helpful step they can take is to fill in a mine after it closes. Replacing trees and ground cover helps natural **habitats** grow back.

Once an ore is removed from the earth, it can never be replaced. But, we can all do our part to make sure metal isn't wasted. Learn about recycling programs in your community. Together, we can protect Earth's **natural resources**.

GLOSSARY

aggregation—a group, body, or mass made up of many distinct parts or individuals.

bonanza—a large and rich mineral deposit.

dense—having a high mass per unit volume.

economic—relating to the production and use of goods and services.

element—any of the more than 100 simple substances made of atoms of only one kind.

environment—all the surroundings that affect the growth and well-being of a living thing.

fortified—having added vitamins and minerals that increase a food's health benefits.

gangue—worthless minerals mixed in with valuable minerals.

habitat—a place where a living thing is naturally found.

magma—melted rock beneath Earth's surface.

mineral—a nonliving solid with a specific chemical makeup that occurs naturally on Earth.

natural resource—a material found in nature that is useful or necessary to life. Water, forests, and minerals are examples of natural resources.

placer—a deposit formed when small particles of valuable minerals have been mixed with sand carried by water or a glacier.

satellite—a manufactured object that orbits Earth. It relays scientific information back to Earth.

SAYING IT

bauxite—BAWK-site

gangue—GANG

garnierite—GAHR-nee-uh-rite

ilmenite—IHL-muh-nite

manganese—MANG-guh-neez

placer—PLA-suhr

rhodochrosite—roh-duh-KROH-site

telluride—TEHL-yuh-ride

Witwatersrand—WIHT-waw-tuhrz-rand

ONLINE RESOURCES

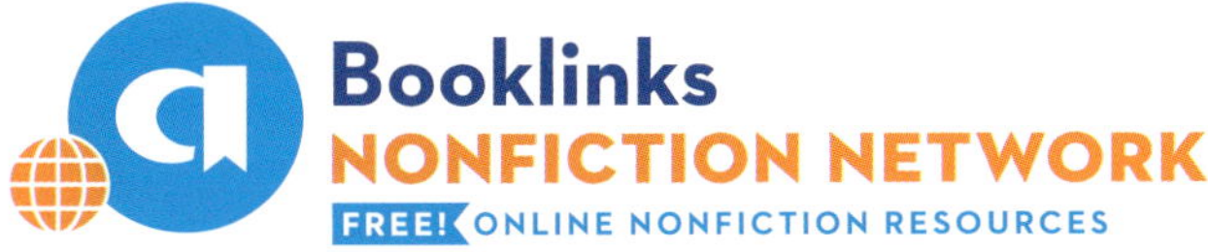

To learn more about ores, please visit **abdobooklinks.com** or scan this QR code. These links are routinely monitored and updated to provide the most current information available.

INDEX